Charles Cole Hine

Fire Insurance

Book of forms containing forms of policies endorsements, certificates, and other valuable matter, for the use of agents and others

Charles Cole Hine

Fire Insurance
Book of forms containing forms of policies endorsements, certificates, and other valuable matter, for the use of agents and others

ISBN/EAN: 9783337256296

Printed in Europe, USA, Canada, Australia, Japan

Cover: Foto ©berggeist007 / pixelio.de

FIRE INSURANCE.

A BOOK OF FORMS

CONTAINING

FORMS OF POLICIES, ENDORSEMENTS, CERTIFICATES,

AND OTHER VALUABLE MATTER,

FOR THE USE OF AGENTS AND OTHERS,

BY

C. C. HINE.

SIXTH THOUSAND.

ENLARGED AND REVISED.

New-York:

SACKETT & MACKAY, STATIONERS AND PRINTERS, 27 PINE STREET.

1866.

Contents.

DWELLING.

MERCANTILE.

SPECIAL.

LEASE, MORTGAGE, &c.

ENDORSEMENTS, &c.

NOTE.—Several changes and ten additions have been made in the recent revision, but, to avoid confusion in reference, the former numbers have not been disturbed, and the fractions 18¼, 18½, 37½, 38½, 45½, 57½, 63½ 72¼, 72½ and 80½ have been used to designate the new items introduced.

ERRATA.

No. 50½. Insurance of Express Co. or Common Carrier. | No. 51½. Mechanics' Lien.

POLICY WRITING—GENERAL INSTRUCTIONS.

[PLEASE READ CAREFULLY.]

I. Clearness, conciseness and completeness should characterize every Policy written. Let there be no uncertain meanings—no useless phrases nor omission of vital points. Say what you mean, and say it in such words as will be fairly and easily understood.

II. Be familiar with pages 72 to 79 of the Instruction Book.

III. Never mutilate, waive or change the printed conditions of the Policy, under any circumstances.

IV. Avoid all ambiguous and general terms, such as "stock in trade," "property," "live stock." If the Policy is on Dry Goods or Groceries—or Druggists' merchandise—or Household Furniture—or Horses—call the property by name, so that it may be certainly known what is insured and what is not.

V. Always give, both in the Policy and your report of it, the CONSTRUCTION, brick, frame or stone—the ROOF, slate, metal, tile, shingles or composition—the NUMBER OF STORIES—the LOCATION, by street, lot or number—the OCCUPANCY—and (if a Mill) the POWER, water, steam or animal.

VI. Write a specific amount on each subject insured, and never cover two or more items in one sum. (See Instruction Book, pages 73 and 77.) *It is better not to do business than to do it under Blanket Policies.* At the home office we practice upon this rule as an imperative one, and cannot encourage in Agents what we do not permit in ourselves.

VII. The following forms are full and varied, and if they do not meet every possible case arising in an extended practice, they will be suggestive, and must be improved upon to suit occasions.

Forms of Policies.

DWELLINGS.

[See pages 8-11 of Instruction Book for Rates and extended remarks on Dwellings.]

No. 1. SUBURBAN DWELLING.

Eighteen Thousand Dollars,

As follows: Ten thousand dollars on his two-and-a-half-story, stone, slate-roofed villa, occupied by the assured as a family residence and situated on the north side of Euclid Street between Day and Martin, Cleveland, Ohio. Three thousand on household and kitchen furniture; one thousand on printed books, pictures and plate; one thousand on wearing apparel; and five hundred on music and musical instruments, all contained therein. One thousand dollars on his brick, slated stable, ten rods in the rear of said dwelling; one hundred and fifty each on his three horses; five hundred on vehicles, harness, whips and robes, and one hundred and fifty on hay, grain and feed therein; and four hundred dollars on the outhouses, fences and shrubbery of said premises. Special reference being had to the assured's application, No. 1, which is his warranty and a part hereof.

No. 2. VILLAGE PROPERTY.

Three Thousand Dollars,

As follows: Two thousand dollars on his brick, shingle-roofed dwelling, occupied by the assured as a family residence and situated on lot two of block forty-three, in the village of Greenfield, Hancock County, Indiana. Five hundred on the household and kitchen furniture; three hundred on the family wearing apparel; and two hundred on provisions and stores, all contained in said dwelling. Special reference being had to assured's application, No. 2, which is his warranty and a part hereof.

Privilege granted to use burning fluid for lights.

No. 3. COUNTRY PROPERTY.

Three Thousand Dollars,

As follows: Fifteen hundred on his two-story frame dwelling house, occupied by the assured as a family residence; three hundred on his household and kitchen furniture; two hundred on his family clothing therein contained; seven hundred on his frame barn, situated twenty rods north of said dwelling, and three hundred on the hay and grain therein contained. All being located one mile west of Fairfield, on the south side of the road leading to Brookville, in Jefferson County, Iowa. Special reference being had to assured's application, No. 3, which is his warranty and a part hereof.

Note.—It is customary in all Policies to recapitulate the amount, the rate and the premium, as for example:

$2,000 @ 60c.	.	$12 00
1,000 @ 90	.	9 00
		$21 00 Prem.

No. 4. ANOTHER FORM.

Five Thousand One Hundred & Fifty Dollars,

As follows:

$1,750 on his one-and-a-half-story brick dwelling, marked A on Diagram.
500 on his household furniture,
250 on his family clothing,
100 on his provisions and stores,
50 on his melodeon,
} therein contained.
400 on his frame horse barn, marked B on Diagram.
300 on his three horses therein, being $100 on each horse.
200 on his harness, robes, whips, cutter and buggy therein.
600 on his frame farm barn, marked C on Diagram.
1,000 on his hay and grain therein.

All being situated on his farm, the north half of section twenty-three in Jacksonville Township, Morgan County, Illinois, as will more fully appear by the assured's application and diagram, No. 4, which is his warranty and a part hereof.

No. 5. CITY RESIDENCE.

Five Thousand Dollars

On his three-story and basement, stone-front, metal-roofed dwelling house, occupied by a tenant and situated on the northeast corner of Ninth and Mound streets, Cincinnati, Ohio. Special reference being had to assured's application, No. 5, which is his warranty and a part hereof.

No. 6. BLOCK OF BRICK DWELLINGS.

Twelve Thousand Dollars,

As follows: Three thousand dollars each on his four brick, metal-roofed, three-story dwelling houses, known as numbers 94, 96, 98 and 100 Linn Street, Cincinnati, Ohio. To be occupied by tenants as private residences only. Privileged to use kerosene for lights. Special reference is had to assured's application, No. 6, which is his warranty and made part hereof.

No. 7. FURNITURE.

Three Thousand Dollars,

As follows: Seventeen hundred and fifty on his household and kitchen furniture; three hundred on piano and music; fifty on gas fixtures; two hundred and fifty on books and pictures; five hundred on family wearing apparel; and one hundred and fifty on plate and jewelry, all contained in the two-story brick, slate-roofed dwelling house, occupied as a family residence by the assured and situated on lot nine, of block forty-three, in the town of Flint, Michigan. Special reference being had to assured's application, No. 7, which is his warranty and a part hereof.

No. 8. ANOTHER FORM.

Twelve Hundred Dollars,

As follows:

$500 on furniture.
300 on clothing.
250 on piano, and music, and printed books.
150 on jewelry and pictures.

All contained in rooms numbers 128 and 130 in the Oliver House, Toledo, Ohio, and occupied by the assured and his family, as per his application, No. 8, which is his warranty and a part hereof.

No. 9. ROW OF FRAME COTTAGES.

Four Thousand Two Hundred Dollars,

As follows: Seven hundred dollars each on his six frame, one-and-a-half-story cottages, standing twenty-two feet apart, situated on the south side of Mechanic Street, between Preston and Shelby, in Louisville, Kentucky, and known as the Stoddart Row. To be used only for dwelling purposes. Special reference being had to the application and diagram, No. 9, of assured, which is his warranty and a part hereof.

No. 10. FRAME DWELLING AND STORE.

Eighteen Hundred Dollars,

As follows: Seventeen hundred on the two-story frame building, known as 42 Main Street, Massillon, Ohio, owned by the assured and occupied by him as a grocery store and family residence; and one hundred on the store furniture and fixtures therein contained, as per assured's application, No. 10, which is his warranty and a part hereof. Privileged to use camphene and other fluids for lighting, and to sell the same; all lamps to be filled, and all sales to be made by daylight only.

Mercantile.

[See Instruction Book, pages 12-21, for Rates and particular directions.]

No. 11. **BRICK STORE AND STOCK.**

Ten Thousand Dollars,

As follows: Four thousand on his two and one-story brick, shingled store, situated on the south side of Washington Street, three doors west of Maine, in the city of Quincy, Illinois, and occupied by the assured on the ground floor for mercantile purposes, and in the upper story by tenants for offices; five thousand on his stock of general merchandise, consisting chiefly of dry goods; and one thousand on his store furniture and fixtures (including scales and a large Herring safe) therein. Special reference being had to assured's application, No. 11, which is his warranty and a part hereof. $10,000 other insurance on the merchandise.

No. 12. **BLOCK OF BRICK STORES.**

Twenty Thousand Dollars

On the block of four stone and brick, metal-roof, three-story buildings, known as the Taylor Block and numbered 1, 2, 3 and 4, situated on the south side of Main Street, between Canal and Fulton, in the city of Madison, Wisconsin, owned by the assured, and occupied for mercantile purposes by tenants. Nos. 2 and 3 having one law office each in the second story, and the Masonic Hall over both in the third. This insurance being in the sum of four thousand dollars each on numbers one and four, and six thousand each on numbers two and three. Special reference being had to the assured's application, No. 12, which is his warranty and a part hereof.

Note.—Never cover two or more buildings, in a block or otherwise, in one sum. If a specific amount on each building is not acceptable to the owner, decline the risk.

No. 13. DRUGS.

[See Instruction Book, page 34, for Drug Stores.]

Five Thousand Dollars

On his wholesale stock of drugs, paints, oils, glass, fancy goods and other druggists' merchandise, contained in the three-story brick, gravel-roofed building, No. 70, on the east side of Commerce Street, Montgomery, Ala., occupied solely by the assured as a drug store. Privilege granted to keep, use and sell burning fluids and kerosene, but not more than three barrels of either shall be on the premises at one time. Special reference being made to his application, No. 13, which is his warranty and a part hereof.

$10,000 other insurance in the Citizens' and Commercial Ins. Cos.

No. 14. STOCK OF LIQUORS.

[See Instruction Book, page 39.]

Four Thousand Five Hundred Dollars,

As follows: Three thousand on his stock of wines and liquors in packages, contained in the three-story brick, shingled store, No. 324, on the east side of Clark Street, Chicago, Ills., and fifteen hundred on his tubs, pipes, pumps and other appliances used for rectifying, contained in the second and third stories of said building. Privilege granted to rectify, without the use of fire heat, not to exceed twelve barrels per day, and warranted by the assured that no smoking or any light or fire be permitted in said building above the first story. Special reference being made to his application, No. 14, which is his warranty and a part hereof.

$4,500 other insurance similarly divided.

No. 15. WHOLESALE GROCER.

Ten Thousand Dollars

On his wholesale stock of groceries, and other merchandise not more hazardous, usual to a wholesale grocer's stock, contained in the three-story brick, metal-roofed store, No. 6 Parson's Block, on Columbia Street, Buffalo, N. Y., occupied by the assured, as per his application, No. 15, which is his warranty and part hereof.

Consent to twenty thousand dollars other insurance.

No. 16. MERCHANDISE GENERALLY.

Seven Thousand Five Hundred Dollars

On his stock of dry goods, clothing, shoes, hats and such other merchandise as is usually kept in country stores, contained in the assured's three-story brick, metal-roofed building, situated on the N. E. corner of Main and Russell Streets, Lexington, Ky., and occupied solely by him for mercantile purposes, as per his application, No. 16, which is his warranty and a part hereof. Privileged to store like goods for other parties, but they are not covered by this Policy. $5,000 other insurance.

NOTE.—Never use the terms "such other *articles*," or "such other *things*," but always say *merchandise* when it is necessary to use this phraseology. We do not favor its use when a more specific can be substituted. Never include "goods on Storage" in an ordinary merchandise policy. See Form 22.

No. 17. ANOTHER FORM.

Ten Thousand Dollars

On grain, flour, feed, salt, pelts, and other merchanise not more hazardous, his own, or held by him in trust or on commission, or sold and not delivered, contained in the three and one-story stone, slate-roofed store house, known as No. 247 Olive Street, St. Louis, Mo., and occupied solely by the assured for a grain and produce store, as per his application, No. 17, which is his warranty and a part hereof. $40,000 other insurance permitted.

NOTE.—Only use form No. 17 when some peculiarity of the applicant's trade renders it necessary. Model merchandise policies upon Nos. 13, 14, 15. See note to No. 20 touching the trust clause.

On no consideration will you use the words "sold but not REMOVED." Such wording would make the insurance follow the property through the hands of a dozen owners if it remained in the same building.

No. 18. OPEN POLICY.

By this Policy of Insurance, the............INSURANCE COMPANY,

In consideration of the premiums endorsed hereon or in the book hereto attached,

DO INSURE..

AGAINST LOSS OR DAMAGE BY FIRE TO THE AMOUNT OF

Such sums, on such property, for such periods and in such places as shall be entered herein with the other particulars of the risk, by the authorized agent of said Company at..................

[OVER.]

Each entry shall apply and have reference only to itself, and all shall be subject to the printed terms and conditions of this policy, and it shall be optional with said agent to make, or refuse to make, any endorsement.

NOTE.—Endorsements on Open Policies must be entered *at once* in the Record. Any delay involves the liability of their being forgotten.

Most Companies have Open Policies printed and books ruled and headed for use.

No. 18¼. FIRE RISK ON VESSEL.

[See Instruction Book, pages 26, 27.]

Five Thousand Dollars

On the side-wheel, high-pressure packet steamer "Major Anderson," including her hull, engines, boilers, tackle, equipments and furniture on board; with the privilege of navigating the Ohio River and the Mississippi River between St. Louis and Memphis.

It is specially understood and agreed that this policy constitutes a fire insurance only, governed by its own terms and conditions; that claims for loss shall be limited to the actual damage to the subject by fire, and no other claim whatsoever shall be allowed under it.

No. 18½. BUILDERS' OR CARPENTERS' RISK.

[See Instruction Book, page 28.]

Three Thousand Dollars

Builders' risk for two months on the two-story frame dwelling house in course of erection situated on Lot 6, Block 42, of the town of Newburyport, Mass. No fire to be allowed in said building except for heating glue in the kitchen fire-place. Doors and windows to be closed and locked at night, and shavings to be thoroughly cleaned out once a week.

NOTE.—Houses erecting increase rapidly in value, and it is customary to insure for short terms and increase the amount on renewal month by month. (See Endorsement No. 72½). Attempts have been made to issue an *increasing* policy to meet this case, but the results have proved to be more novel than practical.

No. 19. **GRAIN ELEVATOR.**

[See Instruction Book, page 18.]

Four Thousand Dollars,

As follows: Three thousand on his grain-elevator building, built of wood, standing detached over one hundred feet, on the west bank of the Sandusky River, north of Fremont, Sandusky County, Ohio, and occupied by the assured solely for elevating, storing and shipping grain. Five hundred on the gearing, belting and other machinery therein, and five hundred on the boiler and engine contained in the brick, gravel-roofed boiler house adjoining. Special reference being had to his application, No. 19, which is his warranty and a part hereof.

$8,000 other insurance permitted.

Note.—Where short risks on grain are to be had, give them the preference, and write small amounts (or none at all) on the building, as may be necessary, to accommodate the short business.

No. 20. **COMMISSION MERCHANT'S POLICY.**

Ten Thousand Dollars,

On grain, flour and meats: his own or held by him in trust or on commission, or sold but not delivered, contained in the four-story stone, slate-roofed building, No. 65 Walnut Street, Cincinnati, Ohio, and occupied solely by the assured for a produce and commission store. Privileged to keep ship stuff and other feed, hay excepted. Special reference being had to the assured's application, No. 20, which is his warranty and a part hereof.

And it is specially understood and agreed that trust and commission goods are covered only to the extent of the actual interest of the assured therein, in advances, debts, liability or otherwise.

Note.—Always put this last clause in policies covering goods "their own, or held by them in trust or on commission, or sold but not delivered" (but never say "*removed*"), and in all other policies covering in a similar manner, and do not insert "for whom it may concern" in a fire policy. See Instruction Book, page 92.

Goods "on storage" not to be covered other than by Form 22.

No. 21. **ANOTHER FORM.**

Merchandise generally (or merchandise without exception), his own, or held by him in trust or on commission, or sold but not delivered. Contained, etc.

NOTE.—Where it is possible, specify the KIND of property covered, and never issue Form 21 when avoidable. It is given rather as a sample of loose practice than for imitation. If used, obtain the rate for the most hazardous goods. Both 20 and 21 should pay full rates, because of the "Blanket" element they contain—covering, perhaps, several different persons or interests for one premium.

No. 22. **WAREHOUSEMAN'S POLICY.**

[See Instruction Book, pages 17, 18.]

Four Thousand Dollars

On goods, wares, merchandise, and all articles held by him on storage and contained in his three-story frame warehouse, standing one hundred feet detached on the south bank of the Miami Canal, between Sixth and Seventh Streets, in Dayton, Ohio, as per his application, No. 22, which is his warranty and a part hereof.

It is expressly understood and agreed that this policy is to cover only the interest of the assured in said property, as the same may be at the time of loss, in advances, charges, commissions or otherwise; and not the interests of owners, consignors or others.

NOTE.—Goods *on storage* are never to be included in an ordinary policy covering merchandise.

Special and Miscellaneous.

No. 23. FLOUR MILL.

[See Instruction Book, page 36, for extended remarks on Flour Mills.]

Five Thousand Five Hundred Dollars,

As follows: Two thousand on their four-story stone and brick, shingle-roofed, water-power flouring-mill building, known as the "Star Mills," standing detached two hundred feet, on the east bank of the Scioto River, five miles south of Chillicothe, in Ross County, Ohio. One thousand on the fixed machinery and gearing, three hundred on the water wheel, and two hundred on scales, trucks, safe, and other furniture and fixtures; five hundred on cooperage, and fifteen hundred on grain, flour and offals, all contained in said mill, and the property of assured, as per their application, No. 23, which is their warranty and a part hereof.

$10,000 other insurance.

Endorsement.—In consideration of $7.50 additional premium, the further sum of one thousand dollars is hereby insured, for sixty days, on one thousand bushels of wheat and its products, belonging to Betts & Jones, of Cincinnati, Ohio.

NOTE.—Do not entertain applications on mills so remote that a personal inspection by yourself is impracticable. See pages 36 and 37 of Instruction Book.

No. 24. HOTEL.

[See Instruction Book, page 38, for Hotels and Taverns.]

Seven Thousand Five Hundred Dollars,

As follows: On the four-story brick, slated hotel building, known as the Empire House, situated on the northwest corner of Main and Canal Streets, Akron, Ohio, and occupied for hotel purposes, five thousand. On the hotel furniture therein contained, as per his schedule on file with this Company, twenty-five hundred. Special reference being had to his application, No. 24, which, with his schedule of furniture, is his warranty and a portion hereof. Consent to $7,500 similar insurance.

NOTE.—Insurance on HOTEL FURNITURE should be made upon a schedule furnished by the applicant. Charge for furniture the same rate as for building. HOTELS *are Special Hazards, and are very unprofitable to Underwriters.* Only the best are desired and at the fullest rates.

No. 25. LIVERY STABLE.

[See Instruction Book, page 40.]

Six Thousand Dollars,

As follows: Two thousand on the brick, shingle-roofed livery stable, and carriage sheds attached, known as the "Fashion Stables," situated on the east side of Parker Street, between Drake and Monroe Streets, in the city of Memphis, Tennessee; fourteen hundred on seven span of carriage horses, being one hundred on each horse; five hundred on the bay trotting mare "Dolly," and five hundred on the stallion "Romeo." Six hundred on four double carriages, being $150 on each; five hundred on five single buggies, being $100 on each; and five hundred on hay, grain and feed—all the property of assured and contained in said Fashion Stable, as per his application and diagram, No. 25, which is his warranty and a part hereof. Warranted by assured that no fire or light, except in close glass lanterns, and that no smoking shall be allowed inside said stable.

Note.—Under no circumstances will you write in one sum on several horses or carriages. *Specify an amount on each in every instance.* If particular animals are too valuable to come under an average sum, name and describe such, and insure agreed amounts on them separately. If this plan is objectionable to the owner and cannot be worked, decline the risk; we cannot suffer "Blanket" insurance in Livery Stables.

No. 26. BREWERY.

[See Instruction Book, page 27.]

Four Thousand Dollars,

As follows: One thousand on his brick and frame Brewery building, standing detached, on out lot, No. 27, in the N. W. suburb of Fort Wayne, Allen Co., Indiana, occupied and used by assured solely for making beer. Five hundred on the kettles, pipes and other machinery, tools and fixtures; and fifteen hundred on the stock of malt, barley and other grain—all the property of assured, and contained in said Brewery. Also, five hundred on his stock of beer contained in the "beer cellar" in the side hill, fifteen feet south; and five hundred on the malt-kiln building attached to the N. E. corner of said brewery, as per application and diagram, No. 26, of assured, which is his warranty and a part hereof.

$6,000 other insurance permitted.

***Endorsement.*—In consideration of $10 additional annual premium, permission is hereby granted for the erection and use of a small still in the S. W. corner of the kettle room, on the ground, no wood work to be within ten feet of the same.**

No. 27. **LUMBER YARD.**

[See Instruction Book, page 40, for Lumber and Wood Yards.]

Six Thousand Dollars

On his stock of lumber, shingles and lath, contained on the premises known as Jones & Wilson's Lumber Yard, lying between the canal and river west of Division Street, in Chicago, Illinois; being three thousand on said stock, located between Division Street and Lime Alley; two thousand between Lime Alley and Boggs Street, and one thousand on dressed flooring contained in the shed adjoining their office near the river. Special reference being had to the assured's application and diagram, No. 27, which is their warranty and a part hereof.

$12,000 other insurance permitted.

NOTE.—Lumber and wood yards are frequently divided by streets or alleys into two or more distinct portions. Specify a separate sum on each in every case, and *decline to issue a policy without such specifications.*

No. 28. **PORK-HOUSE STOCK.**

[See Instruction Book, pages 45, 46.]

Ten Thousand Dollars,

As follows: Seven thousand on bulked and barreled pork, lard and hams; and three thousand on salt and cooperage, belonging to assured, and contained in the three-story brick, slated pork house of Jones & Co., on Sycamore Street (No. 357), between Eighth and Ninth Streets, Cincinnati, Ohio, as per his application, No. 28, which is his warranty and a part hereof.

Endorsement.—In consideration of $50 additional premium, the privilege of smoking meat for one month (ending March 27th, noon) is hereby granted.

NOTE.—The terms "*hogs and hog product*" are often used to cover pork-house stock, and are entirely proper; but under no circumstances should such a stock be covered as "merchandise," or "merchandise generally," nor should "such other articles as are usually kept," &c., ever be introduced; such terms are indefinite and loose, and, in case of loss, may involve great difficulty in the settlement. Always put separate amounts on the meat; the salt and cooperage; the tools and fixtures; the tanks and machinery; the boiler and engine; or the building, as the case may be.

No. 29. WAGON AND BLACKSMITH SHOP.

[See Instruction Book, pages 26, 30.]

Three Thousand Dollars,

As follows: Eight hundred on their frame wagon shop, situated on the north side of Wood Street, between First and Second, Cleveland, Ohio, and occupied by the assured for wagon and plow making, by hand work only. Two hundred on their tools and implements, and six hundred on their stock of materials, and work finished and unfinished, contained therein. Also, three hundred on their brick, shingle-roofed blacksmith shop, standing thirty feet north from said wagon shop; five hundred on their anvils, bellows and other tools and fixtures, and six hundred on their stock of iron and steel, and work finished and unfinished, contained therein; as per their application and diagram, No. 29, which is their warranty and a part hereof.

Warranted by assured that shavings shall be cleaned out of the wood shop every day, and that glue shall not be heated therein during the summer months.

No. 30. RAILROAD PROPERTY.

[See Instruction Book, page 49.]

Let the construction be similar to No. 31, when the Policy is on machine shops and other buildings. If the risk includes a great number of buildings, depots, bridges, etc., call each by name and specify an amount on or in each, as the case may be.

Note.—Policies have been written, "On property of every description belonging to said company, or for which they may be liable, situated on the line of their road," etc. This includes road, bridges, wood, buildings and machinery of every sort, cars, engines, freights, and every other item *in one sum!* Rate of premium is from twelve to twenty per cent., and losses of $500 or less are not to be paid. There is plausibility in the plan, and some companies have statistics that seem to show a profit. But the proportion of insurance to the whole value is very small, being usually $50,000 to $100,000 insurance on five to ten millions of value—*just enough to catch all the losses*, and making the rate, *in reality*, one-tenth to one-fifth of one per cent.—about one-quarter of the rate on choice detached dwellings! The plan is deceptive, and you will decline all such insurances unconditionally, unless specially authorized by the Company after full correspondence.

No. 31. AGRICULTURAL AND MACHINE WORKS.

Ten Thousand Dollars

On their machine works, situated south of the railroad, in Syracuse, N.Y. and occupied by them for the manufacture of agricultural implements, as follows:

$1,000 on their three-story brick, slated building, marked A on Plan.
1,000 " materials, and work finished and unfinished, contained therein.
500 " machinery and tools therein.
400 " one-story brick, metal-roofed engine house, marked B on Plan.
500 " boiler and engine contained therein.
500 " one-story brick, shingled blacksmith shop, marked C on Plan.
200 " tools, machinery, and implements therein.
500 " two-story frame paint shop, marked D on Plan.
1,500 " materials, and work finished and unfinished therein.
1,000 " two-story brick, slated warehouse, marked E on Plan.
3,000 " stock of finished work therein.

Said plan, with the application of assured, numbered 31, which are their warranties and parts hereof, being on file with this Company, as the basis of this insurance.

$20,000 similar insurance elsewhere.

No. 32. OIL REFINERY AND YARDS.

Model the Policy upon No. 31, *after having obtained the consent of the Company to issue it.* (See Instruction Book, page 42.) Avoid the "Blanket" feature carefully, and make the specifications as minute as possible. Oil risks are not favorites and will seldom be taken, and when accepted the fullest rates of premium will be charged.

THE SIX FOLLOWING

Are actual copies from policies on celebrated manufactories. Where the values are immense and the amount of insurance large, it is very necessary to have the policies concurrent; and, to that end, a good and appropriate form is adopted, *printed*, and then either copied or pasted upon the policies of the various companies that insure. These are from such printed forms, and each is a model in its way.

No. 33. **MANCHESTER PRINT WORKS,**

MANCHESTER, N. H.

Twenty Thousand Dollars.

As follows:

$3,125	on	the building, fixed machinery and fixtures therein of the northerly division of their factory building, No. 1, built of brick, situate in Manchester, N. H.
2,375	"	movable machinery, apparatus, tools and furniture therein.
2,000	"	stock therein, raw, wrought and in process, and on supplies for the use of the mill.
3,125	"	the building, fixed machinery and fixtures therein of the southerly division of said No. 1 mill.
2,375	"	movable machinery, apparatus, tools and furniture therein.
2,000	"	stock therein, raw, wrought and in process, and on supplies for the use of the mill.
1,500	"	picker building, fixed machinery and fixtures.
2,000	"	movable machinery, apparatus, tools and furniture therein.
1,500	"	stock therein, raw, wrought and in process, and on supplies for the use of the mill.
20,000	Total.	

Loss, if any, payable to the Treasurer.

With liberty to run such portions of the machinery extra hours, as may be necessary to equalize the work; a watchman to inspect each room where the machinery is not in operation, while the machinery is at work extra hours in other rooms.

Other insurance permitted without notice until required.

No. 34. COLT'S PATENT FIRE-ARM MANUFACTURING CO.,

HARTFORD, CONN.

Twenty Thousand Dollars,

As follows:

$2,370 38	on	their brick and stone armory building, and fixed machinery therein (including the main shafting, main belts, and piping), situate on Huyshope Avenue, and marked B.
4,444 44	"	movable machinery and tools, counter shafting and belting, soda tubs, work boxes, and portable benches contained therein.
1,481 48	"	materials in process of manufacture, and on finished work contained therein.
1,481 48	"	their brick and stone building, connecting with and running East from building B, and fixed machinery therein (including main shafting, main belts and piping), marked C.
2,222 22	"	movable machinery and tools, counter shafting and belting, soda tubs, work boxes, and portable benches contained therein.
2,296 30	"	materials in process of manufacture, on finished work, and on patterns for castings for machinery and tools, and on hardware in store room, contained in said building.
118 52	"	blacksmith shop (including forges, wind pipes and valves, and shafting and belting therein), marked D.
207 40	"	machinery and tools therein.
118 52	"	blacksmith shop (including forges, wind pipes and valves, and shafting, belting and furnaces therein), marked E.
296 30	"	machinery and tools therein.
118 52	"	blacksmith shop (including forges, wind pipes and valves, and shafting, belting and furnaces therein), marked F.
29 64	"	machinery and tools, molding flasks, and patterns therein.
118 52	"	blacksmith and rolling mill shop (including forges, wind pipes and valves, furnaces, shafting and belting therein), marked G.
237 02	"	machinery and tools therein.
44 44	"	blacksmith, hardening, tempering and annealing shop (including forges, wind pipes and valves, furnaces and piping therein), marked H.

[OVER.]

$ 14 82	on tools and materials in process of construction therein.
148 14	" two horizontal engines in building B.
44 44	" one horizontal engine in building G.
88 90	" boiler house connecting with buildings C and F, boilers and connections, and steam feed pump.
88 90	" boiler house connecting with building C, between B and G, boilers and connections, and steam feed pump.
296 30	" office building of brick, marked K.
44 44	" office furniture, fixtures, and iron safes therein.
59 26	" their planing mill of brick, located on Sequassen Street, marked L.
74 06	" machinery, tools, piping, belting and shafting contained therein.
237 04	" their building of brick, located near Charter Oak Avenue, and known as the Tobacco Warehouse.
592 60	" materials in process of manufacture and finished parts contained therein.
1,244 44	" machinery, tools, and steam pipe stored therein.
1,481 48	" finished arms contained therein.
20,000 00	Total.

Permission is given to keep and use Paraffine and Mecca Oils for Lubricating purposes; also, to make necessary repairs, without prejudice to this policy. Other insurance permitted without notice until required.

Liberty given to run until nine o'clock P. M., without prejudice to this policy.

Sharp's Rifle Manufacturing Company's Policy is similar in construction.

No. 35. **WAMSUTTA COTTON MILLS,**

NEW BEDFORD, MASS.

Ten Thousand Dollars,

As follows:

$1,820	on the main building of their Cotton Factory, built of Stone, known as Nos. 1 and 2, and on the fixtures and fixed machinery therein, situate in New Bedford, Mass.
6,532	" movable machinery, apparatus, tools and furniture now in said building, or which may be added.

$1,224 on stock raw, wrought and in process, and supplies therein.
28 " north picker building, and on fixed machinery and fixtures.
34 " south " " " " " " "
56 " movable machinery and apparatus in said north picker building.
66 " movable machinery and apparatus in said south picker building.
62 " stock in north picker building.
62 " " south " "
24 " north engine and boiler house.
38 " south " " " "
20 " north steam engine boiler and apparatus.
34 " south " " " " "

With liberty to have other insurance without notice until required.

No. 36. ASSABET MANUFACTURING CO., SUDBURY, MASS.

Twenty Thousand Dollars

On the several buildings of their woolen factory, built of brick, and on the contents of said buildings, situate in Sudbury, Massachusetts, divided on the several buildings and contents, to apply as follows, viz.:

	On the building, and on all fixed machinery and fixtures therein.	On movable machinery, apparatus, tools and furniture therein.	On stock, raw, wrought and in process, and on all supplies for the use of the factory therein.	Total.
Main factory building	$2,820	$6,350	$3,528	$12,698
West-wing building	282	776	530	1,588
East-wing building	704	952	1,058	2,714
Picker and boiler-house building	212	106	1,058	1,376
Repair shop and wool-drying building	140	140	888	1,168
Dye and wash-house building	104		352	456
				$20,000

With liberty to have other insurance without notice until requested.

Permission is given to run the machinery extra hours, not later than ten o'clock at night.

No. 37. **PEMBERTON COMPANY,**

LAWRENCE, MASS.

Twenty-eight Thousand One Hundred and Twenty-five Dollars.

$3,125	on	the main building of their cotton and woolen factory, built of brick, situate in Lawrence, Massachusetts, and on the fixed machinery and fixtures therein.
12,500	"	movable machinery, apparatus, tools and furniture therein.
4,375	"	stock therein, raw, wrought and in process, and all supplies for the use of the mill.
313	"	the dry-house and dye-house division of the river building, and on fixed machinery and fixtures therein.
1,250	"	stock therein, raw, wrought and in process, and all supplies for the use of the mill.
312	"	the picker, carding and wool-finishing division of the river building, and on fixed machinery and fixtures therein.
1,250	"	movable machinery, apparatus, tools and furniture therein.
625	"	stock therein, raw, wrought and in process, and all supplies for the use of the mill.
750	"	store house, dye house, weaving rooms, sorting room and coal house, machine shop and napping room, and boiler, buildings, and on fixed machinery and fixtures therein.
625	"	movable machinery, apparatus, tools and furniture therein.
500	"	stock, raw, wrought and in process, in the two weaving rooms of the store-house building.
313	"	counting room, finishing and store rooms, and wool-picker building, and on fixed machinery and fixtures therein.
312	"	movable machinery, apparatus, tools and furniture therein.
1,875	"	stock therein, raw, wrought and in process, and all supplies for the use of the mill.
28,125	Total.	

With liberty to have other insurance without notice until requested; and with liberty to remove machinery and to put in additional machinery.

Permission is given to operate the machinery extra hours, not later than ten o'clock at night, commencing at the usual hours in the morning; a man to be kept in each room where the shafting is running, and the machinery is unemployed.

No. 374. **AULTMAN & COMPANY.**

REAPER, MOWER AND SEPARATOR WORKS, situated in Canton, Stark County, Ohio, as follows:

A. (*Brick*) $........... on Building, $........... on finished and unfinished Machines, on first floor, $........... on Machines, finished and in process of finishing, on second floor, $........... on finished and unfinished Material and Belting, on third floor.

B. (*Brick*) $........... on Building, $........... on Wood-working Machinery, on first floor, and $........... on Machinery and material in process of Manufacture, on the second floor.

C. (*Brick*) $........... on Building, $........... on line Shafting, $........... on Iron-working Machinery, on first floor, and $........... on Machinery on second floor, and $........... on Material in process of Manufacture, on second floor.

D. (*Brick*) $........... on Building, $........... on Engine.

E. (*Brick*) $........... on Building, $........ ... on Boilers.

F. (*Brick*) $........ on Building, $........ on Machinery, $........ on Patterns, on second floor of D, E and F, and $........ on Fans and Pipes.

G. (*Brick*) $............ on Building, $........... on Flasks, and $........... on Patterns.

H. (*Brick*) $......... on building, and $......... on ....

I. I. I. (*Frame*) $......... on Building, $......... on Stores, Hardware and Mower Knives, contained therein.

L. (*Frame*) $........... on Building, $.......... on Machines stored therein.

$......... on New Office, $......... on Office Furniture.

$......... on Stores in Frame Store Room.

$......... on Machines and parts of Machines stored in open Sheds.

$......... on Lumber in Lumber Yard.

$......... on Dry House, and $......... on contents.

Application No...... and Diagram, dated Nov. 1st, 1864, on file in Home Office and made part of this Policy.

Other insurance permitted, without notice, until required.

Rents, Leases, Mortgages, etc.

Under favorable circumstances these may be legitimate subjects of insurance, but should be always referred to the Company for specific instructions. They open a wide field for over-insurance and loss, and must be handled with great caution. Be sure that the property is profitable, or decline to cover these interests.

EXAMPLE.—A rented hotel in a western town or an eastern watering place, on which is a mortgage. The owner can insure the *property* and the *rents;* the lessee can insure his *leasehold interest*, and the mortgagee *his mortgage interest.* If they do so, and the property does not make money, a strong temptation arises to "sell out" to the insurers.

No. 38. **RENT POLICY.**

Three Thousand Dollars

On the rents of his four-story brick, metal-roofed buildings known as the "International Block," being numbers 142, 144 and 146, on the east side of Washington Street, in Memphis, Tenn., and occupied for mercantile and office purposes. Privilege granted for customary repairs, alterations and shelving to meet the wants of tenants. Special reference being had to assured's application, No. 38, which is his warranty and a part hereof.

The conditions of this insurance are, that if said premises, or any part thereof, shall be rendered untenantable by fire so as to cause an actual loss of rents to the assured, this Company shall be liable for such loss of rents, not exceeding the sum insured. And the assured agrees to rebuild or repair said premises in as short a time as the nature of the case will admit. Loss to be computed from the date of the fire, and to cease upon the premises being again rendered tenantable. And in case the assured shall not, for any reason, rebuild or repair, then the loss of rents shall be determined by the time which would have been required for such purpose.

NOTE.—*Indemnity* should be at the root of this as well as all other Contracts of Insurance; therefore, the "*actual loss of rents,*" and *not a specified sum* (which may be more or less than such actual loss), should be the measure of the Company's liability.

"The conditions" are the all-important parts of Rent, Lease and Mortgage Policies, and, of course, you will never issue one without them.

Insurance Rent is entirely aside from insurance on buildings, and has nothing to do with the amount of damage to *property*, but relates solely to *loss of rents.*

No 38½. LEASE POLICY.—APPROVED FORM.

The following form, based upon one in the "Practice of Fire Insurance," by Col. DUCAT, of Chicago, is perhaps as near perfect as any *general* form can be:

A party holds a lease for five years at $500 per year, and the Company grants him a Policy of $2,500 on the following terms:

Two Thousand Five Hundred Dollars

On his interest as lessee, in the two-story brick, shingle-roofed building, situated on the northeast corner of Main and Locust Streets, Harrisburg, Penn., and occupied by the assured as a store and dwelling.

It is specially understood and agreed, that if said building shall be totally destroyed by fire, this Company shall pay the whole amount hereby insured, less a deduction of $41.66 per month for the time that shall have elapsed between the date of this policy and the time of such fire; and in case of such damage by fire as shall, without total destruction, render said building untenantable, this Company shall pay at the rate of $41.66 per month for the time said building shall so be untenantable; but all reasonable diligence shall be used by the parties in interest, other than this Company, to repair such damage; and in no case shall this Company be liable for more than the sum insured, nor for any loss other than said leasehold interest.

No. 39. LEASE POLICY.

Upon their lessee interest in the four-story brick, metal-roofed building, situated on the southeast corner of Main and Sixth Streets, Louisville, Kentucky, and known as the Louisville Hotel. Special reference being had to their application and diagram, No. 39, which are their warranties and portions hereof. Reference being also had to the lease of said premises, executed to them by Josiah Sparks, dated January 1st, 1865, and recorded in Jefferson County Mortgage Records, book E, page 742, this insurance covering loss under said lease only.

If said property shall be totally destroyed or rendered wholly untenantable by fire, the amount of loss shall be at the rate of four thousand dollars per annum (the annual rental named in said lease) while it remains untenantable, and a fair proportion for any part so damaged or destroyed; and all reasonable diligence shall be used by the parties in interest (other than this Company) to repair any damage by fire. And the assured agrees to keep the whole current amount of said annual rentals insured, or in default thereof to bear a *pro rata* share of loss.

NOTE.—The "interest" covered must be limited to the actual amount of rent which the lessee would have to pay, and not extended to cover the value of his business, prospective profits, or any other "interest." These are too indefinite and intangible to be made the subjects of insurance. And it should be well ascertained that a lease is *profitable* before *any* insurance is granted upon it.

No. 40. **LEASE POLICY.—N. Y. FORM.**

On his interest as lessee in the brick, tin-roofed building, No. 264 Bowery, New-York City, occupied as a Mantilla store in the first story, and a dwelling above.

It is a condition of the lease hereby insured, that in case of damage to said premises by fire to any extent that can be repaired, the landlord must repair the same without delay; but in the event of a complete destruction of said property, the lease shall be at an end.

It is understood and agreed that this Company shall be liable only in the event of the complete destruction of said building by fire.

No. 41. **THE FOLLOWING POLICY**

Was recently issued in New-York. Parties had leased the Tontine Hotel at an annual rental of $20,000, and subsequently sub-let the same at $40,000.

On his leasehold interest in the five-story brick building, with tin roof, privileged to be occupied as a hotel, restaurant and stores for hazardous purposes, situate Nos. 427 and 429 Broadway, corner of Howard Street, New-York City. Other insurance permitted to the amount of $75,000.

It is understood and agreed that this insurance is to indemnify the assured against loss of revenue from rents received in excess of amount to be paid by him to Thomas Smith, under a certain lease of above premises, dated the 25th of November, 1864, for five years from 1st May, 1865, at the yearly rent of $20,000. In case of the total destruction of the premises, thereby vitiating the above-mentioned lease, then this Company agrees to pay the whole amount of this insurance, less an amount for the expired part of the year, which amount will be calculated at the rate of $20,000 per annum from October 2d, 1865. In case of the partial destruction of the premises, by which the lease is not impaired, then this Company will pay the amount of rent so lost (not exceeding the rate of $20,000 per annum), in excess of amount paid to Thomas Smith, during the time that the premises are untenantable. It being understood and agreed that all diligence is to be used in repairing said building.

No. 42. TRUSTEE AND LESSEE.

JOHN DOE and RICHARD ROE, Trustees of the First National Bank of Syracuse.

Ten Thousand Dollars,

As follows: Four thousand on the safes, counters, desks and other furniture, fixtures and appurtenances belonging to the First National Bank of Syracuse, contained in the two-story stone, slated building, occupied by said Bank, and situated on the southwest corner of Main and Market Streets, Syracuse, N. Y., and six thousand on the leasehold interest of said Bank in said building, as per their application, No. 42, which is their warranty and a part hereof.

If said building shall be totally destroyed by fire, this Company shall pay the whole amount hereby insured thereon, less a *pro rata* deduction at the rate of twenty-five hundred dollars per annum, from the 1st day of May, 1865, the date of said lease. And the Bank shall keep insured the whole current amount to be paid on said lease, or in default thereof shall bear such loss *pro rata* with the insurers.

No. 43. TRUSTEE POLICY.

DO INSURE J. Smith, W. Brown and S. Jones, Trustees of the Millville High School and Free Academy, against loss or damage by fire to the amount of

Nine Thousand Dollars,

As follows:

$2,000 on the three-story brick, slated Academy building, occupied for educational purposes, and situated on Out Lot 63, southeast of the village of Millville, in Monroe County, New-York.

1,000 " the desks, library, philosophical apparatus, globes and other school furniture and appurtenances therein.

[OVER.]

525 " three pianos therein, being $175 on each.
1,475 " the dormitory furniture and household and kitchen furniture and stores therein.
2,500 " the brick, shingled Chapel building, situate 200 feet east of said Academy, and
500 " the organ therein.

$18,000 similar insurance permitted.

Special reference being had to the application, No. 43, of assured, which is their warranty and a part hereof.

NOTE.—Churches, Colleges and other establishments held in trust as above, may be similarly insured.

No. 44. MORTGAGE POLICY.

Four Thousand Dollars

On his interest as mortgagee in the three-story stone, slated store building, occupied for mercantile purposes, situated on the southeast corner of Grand and Conti Streets, Mobile, Ala., owned and mortgaged to the assured by David Morgan, of St. Louis, Mo. Reference being had to assured's application, No. 44, which is his warranty and a part hereof. In case of loss the assured shall assign to this Company an interest in said mortgage equal to the amount of loss paid.

NOTE.—The above form may be used where the mortgagor is a non-resident or inaccessible, but if mortgagor and mortgagee are both at hand it is better to issue the policy direct to the owner under the following form, and let the mortgagee hold it as his collateral.

No. 45. MORTGAGE POLICY.—ANOTHER FORM.

Four Thousand Dollars

(*Loss, if any, payable to Homer Bigelow, mortgagee*) on his three-story stone, slated store building, known as No. 347, on the north side of Rochester Street, in the city of Buffalo, New-York, and occupied by the assured as a wholesale grocer. Special reference being made to assured's application, No. 45, which is his warranty and a part hereof.

NOTE—Strictly speaking, No. 45 is *not* a Mortgage Policy but is something better, as it protects both owner and creditor, while No. 44 gives the owner no benefit whatever, but a loss under it leaves his debt unpaid. No. 45 is on the *property*, while 44 is on the *debt*. The owner pays the premium on 45—the mortgagee on 44. One is property insurance, the other debt insurance.

Savings Banks, and other large holders of mortgages and mortgage insurance, have obtained some concessions from the companies in regard to the acts of mortgagor, more hazardous occupancy, etc., paying therefor advanced rates. Blanks covering these privileges may be obtained from the Company in particular cases that may be approved as you submit them.

No. 45½. CHATTEL MORTGAGE.

The circumstances that would justify the issuance of a Policy on the interest of the holder of Chattel Mortgage would be very rare indeed, and no such insurance should be granted without specific authority from the Company. The point could be reached, in most desirable cases, by an ordinary Policy on the property, issued to the owner, "Loss, if any, payable to A. B, Mortgagee."

No. 46. INSURANCE OF AN INTEREST.

Four Thousand Dollars

On his undivided one-third of the "Star Mills," standing forty feet east of lock No. 14, Ohio Canal, in Akron, Summit County, Ohio, built of wood, and occupied as a flouring mill (water power) by the assured and others, as per his application and diagram, No. 45, which is his warranty and a part hereof.

Said insurance being as follows: Two thousand dollars on the building and two thousand on the machinery therein contained. Additional insurance allowed, not exceeding $15,000 by all parties in interest.

No. 47. LEGATEE INTEREST.

Two Thousand Dollars

Upon his interest as legatee in the two-story brick, shingled dwelling house, situated, etc.

NOTE.—*After the death of the testator*, and pending the settlement or division of an estate, a legatee would have an insurable interest *in the legator's property actually bequeathed to him, but in no other.* While the testator lives, and the will may be changed, he is merely in expectancy, and an expectant has no insurable interest as such.

No. 48. **THE FOLLOWING POLICY**

Was recently issued to George Presbury, Esq., of Baltimore, Md., and is copied here rather as a curiosity than a sample to be followed; although his interest, under the circumstances, as a former, though retired lessee, was one of great value and a perfectly legitimate subject of insurance.

Ten Thousand Dollars

On his interest in the property known as Willard's Hotel, used and occupied for hotel purposes, and situate on Pennsylvania Avenue, in the city of Washington, D. C.; said interest being the agreement of Sykes & Chadwick to pay said Presbury an annual bonus of $18,000 for a term of years. If said hotel shall be burned or damaged by fire, so as to cause a suspension of business therein, and an actual loss of bonus under said agreement, such actual loss shall be borne by the insurers (not exceeding in amount the sum insured) for the term of such suspension, so caused, at the rate of $18,000 per annum, until the property burned or damaged shall be repaired or replaced and again made tenantable, and no longer.

Other insurance $40,000.

No. 49. **GUARDIAN, ADMINISTRATOR, HEIR, &c.**

Make the policy to A. B., as guardian of the minor children of C. D., upon such and such property, describing it the same as in a policy to owner.

Or to A. B., as guardian of John, Mary and Edward, minor children of C. D.

Issue an administrator's policy to "E. F., as administrator of the estate of G. H.," describing the property as usual.

Or issue the policy "to the estate of G. H." This is very frequently done, and is in some cases the best way.

The interest of a MINOR HEIR must be covered through a guardian, trustee or administrator, or not at all.

An heir, of age, can insure his own interest in the property of decedent; the policy should be issued "upon his individual one-sixth interest" (or as the proportion may be).

An HEIR APPARENT has no insurable interest; he is merely an expectant.

No. 50. **RE-INSURANCE.**

[See Instruction Book, page 75.]

NOTE.—Re-insurance at Agencies is not advisable. Do not accept lines in excess of the Company's regulations (see Instruction Book, pagė 21), and do not *grant* a policy of re-insurance to any but the best companies, and then only when you know they retain full lines on the same property, and not then if you can get the risk from the owner direct.

Proper form of Policy is as follows :

Do re-insure the........................ Insurance Company of.................... to the amount of

Five Thousand Dollars

On property covered by its Policy No............... issued at........................ This re-insurance is subject to the same specifications, terms and conditions as the policy which it re-insures, but shall be liable, *pro rata* only, for such proportion (or proportions) of loss as the sum (or sums) hereby re-insured bears (or bear) to the whole amount covered by said policy.

No. 50½. **INSURANCE OF EXPRESS CO. OR COMMON CARRIER.**

Fire policies have frequently been issued to carriers to cover their liability as such. The transaction is not, however, free from obscurity, as the carrier (in his receipt for goods, or bill of lading) almost invariably rejects loss by fire, and express and despatch companies employ *other* carriers (railroad companies, etc.), on whom they have recourse. These insurances should be made the subjects of full correspondence with the parent office, *and should never waive the right of subrogation.*

Ten Thousand Dollars

On property their own or held by them as carriers, on which they would be liable for loss or damage by fire, and for the amount of earned freight thereon; said property being only in the warehouse or freight depot of the Little Miami R. R. Co. at Cincinnati, Ohio, situated between Congress and Front Streets, east of Kilgour Street.

NOTE.—A full rate should be obtained for the locality, and on no account will the "blanket" feature be tolerated in such policies.

No. 51. **FLOATING POLICY.**

[See Instruction Book, page 78.]

Ten Thousand Dollars

On hides, indigo and cotton, stored in any United States bonded warehouse, situated in the city of New-York, south of Chambers Street.

It is understood and agreed that claims under this policy shall only be for such proportion of the whole loss as the amount of this insurance bears to the whole value of the property insured.

NOTE.—Never issue such a policy as the above without special authority from headquarters. "The whole value of the property insured" may be a very difficult thing to ascertain; the hazards covered are diverse, and every way the insurer is at a disadvantage under a Floating Policy.

No. 51½. **MECHANICS' LIEN.**

This is almost identical with mortgage insurance. Both indicate that the property is encumbered, and both need close scrutiny. Grant such insurances only to parties of the highest integrity, and on property of known value.

Issue Policy to the Owner,

(*Loss, if any, payable to Stokes and Gardiner, holders of a Mechanic's Lien*) on his three-story brick, metal-roofed store, No. 162, on the north side of Superior Street, Cleveland, Ohio, occupied for mercantile purposes. Special reference being made to the assured's application, No. 51½, which is his warranty and a part hereof.

No. 52. **BLANKET POLICY.**

(OTHERWISE CALLED THE FOOL'S POLICY.)

[See Instruction Book, page 77.]

Four Thousand Dollars

On his dwelling and barn and the contents thereof including his grist mill and machinery and stock therein for account of whom it may concern with the privilege of ten kegs of powder and twenty of benzine and to use or vend the same by candle-light provided the candle is not put more than one foot inside the kegs his own or belonging to any of his neighbors including damage by lightning thieves and rebels with the privilege of all the insurance he can get and more too and to run all night at discretion with such other privileges as he may have asked or intended to ask for without additional expense.

NOTE.—No note is necessary.

No. 53. **VALUED POLICY.**

A "valued policy" is simply one in which the value of the insured property is inserted,—"Valued at seven thousand dollars" for example; admissible in marine, *but never in fire policies*, as it precludes all evidence or proof (except in fraud) in regard to the value. If you should insert in, say a $6,000 dwelling policy, the words "valued at eight thousand dollars," and after loss it should be proved that the house was worth but $5,000, it would be of no avail, unless fraud were shown; the whole $6,000 must be paid! You see, therefore, the *absolute prohibition which must be placed upon valued policies in fire insurance.*

More forms have been given above than will find their way into ordinary practice, yet there are many other matters on which insurance *might* be made, if it were deemed expedient.

But all *should* not be insured that *can* be. Be in no haste to act upon any proposition for insurance that is out of the usual line, but consult with the Company freely, and only make binding contracts for such when specially authorized to do so.

Endorsements, etc.

I. Enter all Endorsements upon the record at the same time you make them on the policy. Observe this rule as an imperative one, and do not make endorsements except with the record at hand. An hour's delay involves the liability of their being forgotten entirely.

II. Sign and Date all endorsements.

III. Revenue Stamps of five cents each are required on all endorsements that amount to a new agreement or contract. These stamps must always be furnished by the party at whose request and for whose benefit the endorsement is made.

IV. Most endorsements require as much deliberation as an original contract, and you must be entirely clear: 1st, as to the *propriety* of the proposed amendment—that no unfair advantage is sought or granted, or any improper matter "lugged" into the policy; 2d, that no condition of the printed policy is to be waived; 3d, of your own authority to endorse as desired. If doubt exists on any of these points, delay the entry and consult the Company.

No. 54. ASSIGNMENT OF POLICY.

Blank forms for assignment are on the inside of the policy. No assignment is legal without the written consent of the Company, and it is your duty to withhold consent from all assignments that do not meet the full approval of your judgment.

Assignment requires the same revenue stamp as the original policy, to be furnished by the parties and never by the Company.

No. 55. ASSIGNMENT AFTER A FIRE.

While the insured *cannot* assign his policy without the Company's consent, he *can* assign his claim—if he has one—at will, the same as any other debt.

No. 56. ENDORSEMENTS AFTER A FIRE.

Upon no consideration will you make an endorsement upon a policy *after a fire*, before settlement of the loss. Do not give consent to other insurance, assignment of policy, change of occupancy, or any other thing whatever. If the loss is a fair one the party will not suffer, but you may complicate matters and injure all sides by tampering with a policy after a fire.

No. 57. PAYMENT OF PARTIAL LOSS.

In consideration of seventeen hundred dollars paid in full of all claims under this policy for loss or damage by the fire of October 17th, 1865, the same is hereby canceled in that amount, leaving the sum of thirty-three hundred dollars now pending.

To be signed both by assured and the agent; and such other receipts as the printed forms of the Company require, may be taken and forwarded, to complete the files at the parent office.

No. 57½. RESTORING ORIGINAL AMT. AFTER PAYING PARTIAL LOSS.

EXAMPLE.—A $3,000 policy has run four months when a loss of $1,700 occurs. Repairs are made and owner wishes the full amount restored. Rate is one per cent.

In consideration of $12.30 additional premium paid, the sum of seventeen hundred dollars is hereby covered for eight months, to expire with the other insurance; restoring the policy as before the payment of loss.

No. 58. RETURNED PREMIUM ON A CANCELED POLICY.

In consideration of twenty-three $\frac{42}{100}$ dollars returned premium on the within policy, to me paid, said policy is hereby canceled and surrendered to the Company.

To be signed by the insured. All receipts for twenty dollars (or more) money paid, require 2-cent revenue stamps.

No. 59. CANCELLATION OF POLICY.

Sometimes it is sufficient to write the word CANCELED in a bold hand across the face of the policy, but if any money is returned use form No. 58.

No. 60. PARTIAL CANCELMENT TO REDUCE AMOUNT COVERED.

The amount hereby insured is this day reduced to five thousand dollars, and the expiration of this policy extended to July 20th instead of April 5th.

Example.—Suppose a $10,000 annual policy has run 8½ months, the unearned premium for the remaining 3½ months would carry half the amount 7 months, and the above arrangement *retains all the premium* and gives the insured its equivalent in extended time. The plan is a valuable one and should be adopted whenever practicable.

No. 61. RECEIPT FOR LOSS WHEN POLICY IS BURNED OR LOST.

In consideration of $500 to me this day paid by the United States Insurance Company of New-York, for loss by fire under their policy, No. 397, issued to me from the Utica Agency, I hereby receipt in full for all claims whatsoever by reason of said policy, and forever discharge and cancel the same. And I hereby agree to protect and defend said company against all persons or claims on account of said policy, the same having been burned, lost or mislaid.

7-cent revenue stamps required.

No. 62. ABANDONMENT.

By its own express terms there is no abandonment under a fire policy; and, although some large adjustments have, for the convenience of all concerned, been conducted as upon that principle, yet you will never for a moment entertain it or consent to it without express authority from your Company.

No. 63. SUBROGATION.

[See Mortgage Policy, No. 44.]

It is expressly agreed that, in case of loss, the insured shall assign to this Company a portion of said mortgage (or interest or claim, as the case may be) equal to the amount of loss paid.

No. 63½. ASSIGNMENT TO MORTGAGEE.

EXAMPLE.—A party holding a policy, mortgages the insured property and wishes to give the mortgagee his policy as collateral security. If the transaction meets your approval, endorse as follows:

Consent is hereby granted to a mortgage on (or including) the insured property, and loss, if any, shall be payable to A. B., mortgagee, who may hold this policy as collateral security.

No. 64. AVERAGE CLAUSE.

It is understood and agreed that claims under this policy shall only be for such proportion of the whole loss as the amount of this insurance bears to the whole value of the property insured.

For use chiefly in floating or blanket policies.

No. 65. RENEWALS OF POLICY.

These are now never made by endorsement, but by regular "Renewal receipts," furnished by the Company, as *the signatures of the officers are required the same as upon the policy itself.* Do not venture upon endorsement renewals, or you may find them of no legal effect.

No. 66. PRIVILEGE TO ERECT AN EXPOSURE.

In consideration of twenty dollars additional annual premium, consent is given to build another frame store adjoining the above-described property on the south. Carpenter work limited to two months.

No. 67. CHANGE OF OCCUPANCY.

Notice is hereby accepted of the removal of Jones & Co.'s stock of groceries from the insured building, and the occupancy of the same by Brown & Smith, as a Carpet store. Privileged for necessary changes in shelving and counters.

No. 68. PARTIAL CHANGE OF OCCUPANCY.—MORE HAZARDOUS.

In consideration of fifteen dollars additional premium, privilege is granted to occupy the lower story corner room and the cellar as a Drug store.

No. 69. REMOVAL OF GOODS.

Permission is granted to remove the insured merchandise (or household furniture, as the case may be) from the location herein described to the two-story frame building, No. 142 Main Street, corner of State; the risk ceasing at the former location from date of this endorsement.

NOTE.—If to a more hazardous locality collect proper additional premium, and begin the endorsement "In consideration of, etc." If to a less hazardous locality and a reduction of rate is demanded, extend the time as per page 74 of Instruction Book. See also the principle involved in Endorsement No. 60.

No. 70. ADDITIONAL TENANTS.

In consideration of ten dollars additional premium, consent is given for the addition of a Tobacconist on the lower floor and a Tailor shop in the second story.

No. 71. INCREASE OF HAZARD AT RENEWAL.

In consideration of the additional exposures now being erected on the west, the rate is hereby advanced to 1.75 instead of 1.25 as heretofore.

No. 72. DECREASE OF HAZARD AT RENEWAL.

In consideration of the vacation of insured premises by the Drug store above mentioned, the rate is hereby reduced from 1.50 to 1.00.

NOTE.—Endorsements are always to be made upon the policy itself, and never on the renewal receipt.

No. 72¼. INCREASED AMOUNT ON RENEWAL.

On renewal this day, the amount hereby insured is increased from $3,500 to $5,000.

NOTE.—If there are several items covered in the policy state specifically what the increase is on.

N. B. Never make an endorsement upon the renewal receipt, but if any changes are agreed upon, endorse them *on the policy;* note them in the record book and then renew the policy as changed.

No. 72½. DECREASE OF AMOUNT ON RENEWAL.

On renewal this day the amount hereby insured is diminished from three thousand to two thousand dollars.

NOTE.—See NOTE and *N. B.* above.

No. 73. PORK-HOUSE PRIVILEGE.

In consideration of twenty-five dollars additional premium, consent is hereby given to smoke meat and render lard on the above-described premises for one month, ending February 16th, at noon.

No. 74. CARPENTERS' PRIVILEGE.

In consideration of ten dollars extra premium, privilege is hereby granted to re-roof and make other repairs upon and within the building for two months.

Note.—It is customary, where the original rate is satisfactory, to grant fifteen days carpenters' risk on an annual policy without extra charge.

No. 75. TO RUN MACHINERY EXTRA HOURS.

For one hundred dollars additional premium, consent is hereby granted to run the factory all night, with double sets of hands.

Note.—Charge double rates for night work as above. This rule is well established among the companies and is only just. You will observe it in all cases.

No. 76. CANCELLATION OF A PRIVILEGE.

The privilege to run at night is hereby canceled, and thirty-four $\frac{73}{100}$ dollars premium refunded.

Note.—When such cancellation is the subject of prior agreement, or a matter of policy or courtesy with a valued customer, the above will be in order. Generally the margins are all against the insurer, and he should be in no haste to refund premiums. Apply the rule in No. 60 where practicable, instead of returning money.

Retain short rates on above cancellation.

No. 77. ADDITIONAL INSURANCE.

Notice this day accepted of $5,000 other insurance in the American Insurance Company.

Note.—Only consent to additional insurance where you know the values justify it. If a building or stock is getting too fully covered, withhold consent. *Be firm, and you may prevent a fire.*

No. 78. TO KEEP COAL AND BENZINE.

For forty dollars extra premium, consent is given to store coal and benzine within the premises described above, not to exceed twenty barrels of either; and warranted by the insured that no smoking, nor any fire or lights shall be permitted in any room or cellar where said articles are stored.

NOTE.—Storage of benzine is admitted by very prudent insurers. Be posted in regard to the usage of your company and observe it strictly.

No. 79. TO USE OR VEND COAL OIL OR CAMPHENE.

For seven $\frac{50}{100}$ dollars, privilege is granted to vend coal oil and camphene, and to use the same for lights, provided lamps are trimmed and filled by daylight, and no camphene to be drawn or sold at night.

No. 80. TRANSFER OF INSURANCE FROM MERCHANDISE TO BUILDING.

The insured having removed (or sold out) his stock of goods (or household furniture) from the above-described building, belonging to him, the insurance on the goods is hereby canceled and made to attach from this date to said building.

No. 80½. SALE OF BUILDING (OR GOODS) AND TRANSFER OF INSURANCE TO OTHER PROPERTY.

In consideration of the sale of the above-insured property, the insurance thereon is terminated and hereby transferred to the adjoining similar building, No. 227 Walnut Street, subject to all the terms and conditions of this policy.

NOTE.—Generally it will be better to cancel the old and issue a new policy, to meet a greater or less hazard or rate, and in no case should the above endorsement be made unless the two buildings are similar in these two points.

The same remarks will apply to merchandise, unless the transfer is to similar goods in the same building.

No. 81. EXTENSION OF POLICY TO COVER MORE PROPERTY.

Consent is granted that this insurance shall attach to all wheat and corn belonging to assured, contained in the above-named warehouse.

NOTE.—Use the above with caution. If a lot of flour, a *particular lot* of grain, or other property, has been covered, such extension to cover more *like* property, similarly stored and for same parties, will be in order, but if the property is more hazardous, or the ownership different, or the locality other, then issue a new policy and collect the premium.

No. 82. EXTENSION OF TIME.

For three $\frac{50}{100}$ dollars additional premium, this insurance is extended fifteen days, to expire November 22d instead of November 7th.

NOTE—This is practically a Renewal, and should only be used in an emergency.

No. 83. ENDORSEMENTS ON OPEN POLICIES

Should contain all the particulars of the risk as to property covered, location, amount, rate, time, and such other matters as may be pertinent; usually the printed headings of the open policy book call for all the needed entries. They should be at once put upon the record book, as any delay may result in their being overlooked and forgotten.

No. 84. ENDORSEMENTS ON DRAFTS.

Make them payable to the order of the Company and not to any officer (unless otherwise instructed). Particular individuals may be absent, but the "Company" has always a representative at home.

No. 85. BINDING RECEIPTS.

Received of Richard Roe, fifty dollars premium for insurance as per his application, No. 742, to the Reliable Insurance Company of Podunk, which insurance is made binding until rejected by said company and notified to said Roe.

NOTE.—Binding receipts may be issued during a temporary lack of policies, or for such other emergencies *as the company may specifically authorize;* but never for special hazards that you submit for examination. Wait for final advice on these before binding the company.

No. 86. "FOR WHOM IT MAY CONCERN"

Has too much of the blanket element to find favor here. Only when the propriety of covering the goods of several men for one premium, or of one man insuring another man's property is clear, should this term be used.

In Addition

To the foregoing, forms might be given for Certificates of Insurance, Renewals, etc., but as these must bear the signatures of officers, they can only be produced at the parent office.

Loss proofs, Loss receipts, Appraisement blanks, Drafts, Accounts current, Expiration notices, and various other matters, are supplied by the several companies printed in convenient forms, and their introduction here would be superfluous.

It is believed that all the points of ordinary practice are herein covered, and that several matters novel and useful, not heretofore presented, are also furnished; and the hope is indulged that the work may prove a convenience and a benefit to all in the profession, into whose hands it may come.

C. C. HINE.

www.ingramcontent.com/pod-product-compliance
Lightning Source LLC
Chambersburg PA
CBHW021439090426
42739CB00009B/1553
9783337256296